THE AUDUBON SOCIETY POCKET GUIDES

A Chanticleer Press Edition

Peter Katsaros
Member, North American Mycological Association and
Mid-Hudson Mycological Association

FAMILIAR MUSHROOMS

Alfred A. Knopf, New York

This is a Borzoi Book
Published by Alfred A. Knopf, Inc.

Prepared and produced by Chanticleer Press,
New York.
Typeset by Dix Type Inc., Syracuse, New York.
Printed and bound by Dai Nippon, Tokyo, Japan.

First Printing.

Library of Congress Catalog Number: 90-052503
ISBN: 0-679-72984-4

Trademark "Audubon Society" used by publisher under
license from the National Audubon Society, Inc.

Cover photograph: *Amanita muscaria* var. *flavivolvata*
by Harley Barnhart.

Contents

How to Use This Guide

Mushrooms play an integral part in the natural world; yet relatively few people are aware of them, except in a general way. Looking for mushrooms is a rewarding outdoor activity, whether gathering them for the table, photographing or painting them, or studying them scientifically.

Coverage

This new guide covers 80 of the most common, colorful, and interesting mushrooms found in North America, north of Mexico. The range covers the conterminous United States and southern Canada, an area comprising two-thirds of this hemisphere's North Temperate Zone.

Organization

This easy-to-use pocket guide is divided into three parts: introductory essays; illustrated accounts of the mushrooms; and appendices.

Introduction

Three introductory essays will supply you with fundamental information about mushrooms. "Identifying Mushrooms" outlines the key elements one should look for when attempting mushroom identification. "Mushroom Hunting" tells you the basics of when and where to look for mushrooms. Be sure to read "Mushroom Poisoning."

The Mushrooms	This section includes full-color identification photographs of 80 familiar North American mushrooms. Each photo clearly depicts the shape, color, and overall appearance of the featured mushroom; the pictures are arranged in a visual sequence to make recognition easy. Facing the illustrations are the text accounts for each species. These begin with an introductory paragraph that fills the reader in on interesting mushroom lore; next is a description of key identifying features of the species. Following the description is a note on edibility; information and warnings about look-alikes; a description of the species' typical habitat; and an indication of the mushroom's geographical range. A silhouette provides a quick reference to the shape. Poisonous species are identified with the symbol $\otimes$.
Appendices	Following the species accounts and the color illustrations is a "Guide to Families," which outlines 30 mushroom families. Line drawings show you key features that distinguish different kinds of mushrooms, including basic shapes and the areas where the spores (the minute reproductive bodies) are reproduced. A listing of spore colors provides important information for mushroom identification.

Identifying Mushrooms

Mushrooms are nonflowering plants that typically produce large fruiting bodies. In contrast to other types of fungi, such as molds and mildews, mushrooms are normally readily distinguishable by their size, shape, and color. The term "mushroom" was once understood to denote only the familiar supermarket mushroom; today, however, we group together under this one heading all the large, fleshy fungi—whatever their shape, and including the toxic species that were long known as "toadstools."

Fungi have existed for many millions of years, but they are more delicate—more "perishable"—than most other plants and so left few clues to their ancient history. Contemporary mushroom study, called "mycology" (from the Greek "mikat" for mushroom), deals with thousands upon thousands of different species. Even today, new kinds are still being found. Traditionally mushrooms were assigned to different categories based upon their overall appearance. Modern mycologists, however, adopt a different view, grouping mushrooms according to similarity in microscopic structure and chemical affinity.

To make a successful identification of a mushroom, it is crucial to examine its structural features closely. The following guidelines suggest what to look for when examining a mushroom.

Shape, Texture, and Gills Most mushrooms have the same basic "umbrella" shape —a stalk topped by a cap. Within this group, however, there are several easily recognized features that provide clues to a mushroom's identity.

Look at the cap of the mushroom you wish to identify. It may be flat, rounded, bell shaped, or funnel shaped. Its texture may be rough, smooth, or scaly; and the surface may be dry or somewhat sticky. Each of these features is a distinctive field mark that will help you to identify the mushroom.

The underside of the cap is where the spores (minute reproductive bodies) are produced. In many stalked mushrooms with a cap, the underside bears thin flaps of tissue, called *gills*, that radiate from the stalk to the outer edge of the cap, like spokes. (Look at the underside of a supermarket mushroom for an easy illustration.) In most gilled mushrooms—often called "agarics"—the gills are attached firmly to the stalk. A few species have gills that run part of the way down the stalk *(descending gills)*, and a few others have gills that are not attached to the stalk at all *(free gills)*.

Other mushrooms look like agarics when viewed from above, but instead of gills have a flat surface riddled with pores (such as the boletes) or an array of downhanging icicle-like "teeth" (such as the hydnums). When examining mushrooms with stalked caps, it is important to note if there is a ring of tissue surrounding the upper portion of the stalk. The ring may be skirtlike (as in *Amanita*) or cobwebby (as in *Cortinarius*). A membranous sheath of tissue, or its fragmented remains, may be found at the stalk base. **NOTE:** The presence of this basal tissue is almost always an indication the species is an *Amanita* and may be one of the deadly poisonous species.

Color Whatever a mushroom's shape, its color is of primary importance in making identifications. But moisture, sunlight, and other factors may affect a mushroom's color; the pigments in young mushrooms are often more intense than those in aging specimens.

Size Mushrooms vary in size from tiny cup-shaped forms to relatively massive flat caps that can be more than one foot in diameter. Some puffballs are even larger; according to legend, at least, they have been mistaken for sheep in the distance.

Edibility and Toxicity	Estimates vary, but experts agree that a minimum of 5,000 different kinds of mushrooms occur in North America. Most of them are bitter or flavorless. A few, however, are choice edibles, and a somewhat larger number are deemed suitable table fare, depending upon one's taste. A handful of toxic species cause some degree of gastrointestinal upset or psychotropic disturbance; a few are harmless on their own but poisonous when consumed with alcohol. And a few notorious species are fatal if eaten. Although experienced mycologists do often eat the mushrooms they find in the wild, such a practice is not recommended for novices.
Similar Species	Some species have one or more look-alikes, which are briefly discussed under this heading in the text accounts. Note toxic look-alikes in particular.
Habitat and Range	Some mushrooms grow only on decaying wood, while others are found in grassy areas or other specific locales. Where range is given in broad terms, such as "widely distributed," it should be understood that the mushroom may appear in any area with suitable habitat.

Mushroom Hunting

Searching for mushrooms is an enjoyable exercise shared by many enthusiasts. The properly equipped mushroom hunter always carries a collecting basket, a knife for uprooting species, and waxed paper in which to wrap specimens. Mushrooms wrapped in plastic material will deteriorate rapidly.

Season

Relatively few mushrooms will be found in the spring, but that is when the most searched-for mushroom, the Morel, makes its appearance. Note that toxic false morels (genus *Gyromitra*) may be found growing at the same time. Most mushroom species begin to appear in late summer, with the majority occurring in the fall. This is the season to search for most of the species in this book.

Where to Search

Decaying wood is host to a large number of mushroom species, including fine edibles such as the Oyster Mushroom *(Pleurotus ostreatus)* as well as poisonous species such as the Deadly Galerina *(Galerina autumnalis)*. Lawns and open fields, particularly if well-manured, produce their own set of species, including the Fairy Ring Mushroom *(Marasmius oreades)* and the Meadow Mushroom *(Agaricus campestris)*.

Many mushrooms grow only in the vicinity of certain types of trees or other plants. The mushroom is connected by underground strands to the rootlets of the tree, forming a mutually beneficial arrangement known as a mycorrhiza (i.e. fungus root). It is believed that all members of the family *Russulaceae*, for instance, are mycorrhizal. Serious mushroom hunters, therefore, learn to recognize different types of trees.

Mushroom Poisoning

Never eat any wild mushroom unless it has been carefully identified as to both species and edibility. Since mushrooms show significant variation in color, size, and other features, accurate identification is often very difficult. Furthermore, some mushrooms are harmless for most people but cause distressing and unpredictable allergic reactions in a few.

If you do decide to eat wild mushrooms, here are a few rules of thumb: 1) do not eat any *Amanita* species and be especially careful in identifying *Amanita* look-alikes or any other white mushroom; 2) avoid LBMs ("Little Brown Mushrooms") and large brownish mushrooms, especially those with pinkish, brownish, purple-brown, or blackish gills; 3) avoid false morels; 4) never eat any wild mushroom without cooking it first; 5) do not take alcoholic beverages in conjunction with wild mushrooms; and 6) only consume small quantities of any wild fungus.

Some toxic mushrooms cause no more than mildly unpleasant sensations. Others, such as the Jack O'Lantern *(Omphalotus illudens)*, may cause such severe gastrointestinal problems that unwary epicures may be hospitalized. Most dangerous of all are mushrooms in the complexes of species represented by the Destroying Angel *(Amanita virosa)* and the Death Cap *(Amanita phalloides)*. These mushrooms have caused a number of fatalities. Anyone who gathers mushrooms for the table should memorize the field marks of species in this group.

The mushroom shown at left is a Destroying Angel, common in some parts of the United States. It resembles some popular edible species, but there is little chance of mistaking it for an edible mushroom if its identifying features are carefully checked. It is white overall, with gills that do not reach (or just barely reach) the stalk. It has an ample, membranous white ring on its stalk, and a saclike structure at the stalk base, which may be buried in leaves or loose soil. This sac is one of the most important features for identification of this deadly species; and its many equally dangerous allies. Be sure to dig white mushrooms gently from the soil, and search carefully for remnants of a sheathing cup.

Destroying Angel has white, skirtlike ring, stalk, cap, saclike structure at stalk base, and gills. The gills do not reach stalk, or just reach it.

THE MUSHROOMS

Clustered Psathyrella *Psathyrella hydrophila*

More than 400 species of *Psathyrella* occur in North America. Many bear a close resemblance to one another, making microscopic examination essential for identification. Other factors can be useful, as well. The Clustered Psathyrella, for example, grows on decaying wood; occasionally it may spring from buried wood, and thus appear to be growing on the ground.

Identification	Cap 1–2″ wide; broadly curved to flat, and moist. Bright or dull reddish brown, fading to pale tan. Cap edge at first shows torn fragments of veil (irregular white fibers). Gills closely set are brown to dark reddish brown. Narrow, shiny white stalk to 3″ or more.
Edibility	Unknown.
Similar Species	Other similar *Psathyrellas* best differentiated microscopically. *P. septentrionalis* has sharply tooth-shaped veil fragments.
Habitat	Usually clustered on decaying wood of hardwood trees.
Range	Widely distributed throughout North America.

Clustered Collybia *Collybia acervata* ⊗

The scientific name of this species means "a group of coins"—perhaps because a cluster of these mushrooms looks like a handful of small gold coins. Mushroom hunters collecting this species for table fare gamble with their well-being. Some people find it merely bitter, but it has proved toxic to others.

Identification Cap ½–2″ wide; reddish brown with purplish cast when young, fading to light reddish buff or paler. Smooth and dry. Closely set whitish- to pinkish-tinged gills attached to stalk, may be almost free. Stalk to 4″; narrow, ringless, reddish brown, usually darker than cap, with white hairs at base.

Edibility Poisonous.

Similar Species Family Collybia, *Collybia (Clitocybula) familia*, has a buffy white stalk. Fused Marasmius *(Marasmius cohaerens)*, with a dotted stalk, grows on fallen leaves of hardwoods and on humus.

Habitat Clustered on or near rotting conifer logs and stumps.

Range Northeastern, northern, and western North America.

Fly Agaric *Amanita muscaria* var. *muscaria* ⊗

Used at one time to make a fly poison, the Fly Agaric figures prominently in the folklore of peoples in widely separated areas of the world. Formerly believed to be a deadly poisonous species, it is now known to produce states of delirium and raving, but few fatalities. Its potency seems to vary geographically, and several varieties are recognized.

Identification   Cap 2–10″ wide; rich red to reddish orange, striate (lined) at rim, with scattered white patches. Gills white, closely spaced; free of stalk or just touching it. White stalk, 2–7″, ringed near top, with spiraling bands and patches at the base.

Edibility   Poisonous.

Similar Species   A form with an orange to yellow cap, the Yellow-orange Fly Agaric (*A. muscaria* var. *formosa*) occurs most often in eastern North America. Toxic Panther *(A. pantherina)*, brownish with collared basal bulb.

Habitat   On the ground under conifers or hardwoods.

Range   West Coast to Rocky Mountains; rare in the East.

Destroying Angel *Amanita virosa* ⊗

The Destroying Angel belongs to a complex of the most deadly mushrooms in the world, which are responsible for the vast majority of mushroom fatalities. There are some edible *Amanita*s, but the risk of a mistake is always present; no *Amanita* should be eaten.

Identification
Cap from 2–5″ broad; bright white; flat or with a small central bump, with smooth texture; may be slightly sticky. White, closely set gills may be free of the stalk or just reach it. Stalk to 8″ tall; white and cottony, with a skirtlike white ring near top and basal membranous cup.

Edibility
Deadly poisonous.

Similar Species
Several similar deadly *Amanita*s are also known as Destroying Angels. White *A. verna* and *A. bisporigera*, buff *A. ocreata*, and greenish *A. phalloides* best distinguished by spores. Similar *Agaricus* lack basal cup or remnants.

Habitat
Singly or in small scattered clusters; on the ground in coniferous, deciduous, or mixed woods.

Range
Widely distributed throughout North America.

Rooting Collybia *Oudemansiella radicata*

The common name of this very familiar mushroom was retained when the species was moved from the genus *Collybia* to *Oudemansiella*. Research has shown the Rooting Collybia is a variable species belonging to a complex of very similar mushrooms, details of which are still being investigated.

Identification
Cap 1–4½″ wide; brown, gray-brown, or paler; flat or with central bump, usually somewhat sticky and wrinkled from center outward. Gills white, fairly well spaced. Stalk to 8″; slender, ringless; white above and brown elsewhere, appearing twisted; widens at base, continuing into ground with rootlike, pointed extension.

Edibility
Edible.

Similar Species
O. longipes has a velvety, usually dry, cap.

Habitat
On the ground under deciduous trees or in grassy areas.

Range
Eastern North America to Midwest; more common in East.

Fairy Ring Mushroom *Marasmius oreades*

When Fairy Ring Mushrooms grow in unobstructed areas such as lawns, meadows, or even golf courses, they spread evenly in all directions, forming a circle of mushrooms known as a fairy ring. Left undisturbed, such rings may in time become quite large. In dry weather the mushrooms wither, but they spring magically back to life after a rainstorm.

Identification Cap to 2″ wide; bell-shaped or outspread with a central bump; deep brown to light yellow; dry. Variable yellowish-white gills sometimes reach the stalk; may be closely set or not. Narrow, tough stalk, to 3″, is yellowish white or a bit darker.

Edibility Choice, with caution; resembles small toxic species.

Similar Species Poisonous Sweating Mushroom *(Clitocybe dealbata)* is gray-pink with gills usually running down stalk a bit. Toxic *Inocybe* usually have yellow- or gray-brown gills.

Habitat In grassy areas everywhere.

Range Widely distributed in North America.

Deer Mushroom *Pluteus atricapillus*

The Deer Mushroom, sometimes called the Fawn Mushroom or *P. cervinus*, is considered a familiar species. Yet it is frequently confused with others in its genus that resemble it. A white variant also occurs.

Identification — Cap 1½–5″ wide, flat or with a long, low bump, somewhat hairy centrally, and may be a bit sticky. Often somewhat wrinkled, with color varying from dark brown to paler grayish brown. Closely set gills are white when young, pink at maturity, and free of the stalk. Stalk is up to 4½″ long and ½″ thick; firm; white to brown-tinged, with occasional dark fibers.

Edibility — Edible.

Similar Species — Larger *P. magnus* has coarse wrinkles. Black-edged Pluteus *(P. atromarginatus)* has dark-edged gills. Other *Pluteus*, including white species, require microscopic differentiation.

Habitat — Decaying trees, stumps, and sawdust piles.

Range — Widely distributed in North America.

Questionable Stropharia *Stropharia ambigua*

There is no sharp line of demarcation separating this genus from two others, *Naematoloma* and *Psilocybe*; and some experts believe the three groups should be combined. The Questionable Stropharia received that name because its ring is usually poorly formed and sometimes actually disappears altogether—leaving its classification ambiguous.

Identification
Cap 2–5″ broad, smoothly curved to almost flat, and sticky. Yellow at first with small, white flecks and patches that disappear. Gills closely set, grayish at first, becoming purple-brown. Stalk to 6″ long, ½″ thick; white with a disappearing white ring above and white strands at base.

Edibility
Reportedly edible, but lacks flavor.

Similar Species
Amanita and *Armillaria* species have white to yellow gills.

Habitat
On the ground in coniferous or mixed woods.

Range
West Coast from California north; common in the Pacific Northwest.

Deadly Galerina *Galerina autumnalis* ⊗

The common name of the Deadly Galerina speaks for itself. This species contains the same death-dealing toxins that are found in the Destroying Angel (*Amanita virosa*). The Deadly Galerina is a classic example of an extensive group of harmless-looking fungi, collectively known as the LBMs ("Little Brown Mushrooms"). It is the finest argument available for never eating any mushroom of that type.

Identification	Cap from 1–2½″ wide, usually with a small central bump; slightly sticky; dark brown when moist, fading to buff when dry. Usually faintly striate (lined) around the edge. Gills closely set; yellow at first, turning rust-brown later. Stalk to 4″; narrow; whitish above, brown below, with a thin, pale ring.
Edibility	Deadly.
Similar Species	Deadly *G. marginata* differs microscopically. Changing Pholiota *(Pholiota mutabilis)* has scaly stalk.
Habitat	Clustered to scattered on decaying logs of all kinds.
Range	Widely distributed in North America.

Parasol *Lepiota procera*

This stately species has been classed in several genera
—*Macrolepiota*, *Leucoagaricus*, and *Leucocoprinus*—
by a variety of mushroom experts. Despite the
controversy in determining its proper placement, the
Parasol is a relatively easy mushroom to identify in the
field, and is a great favorite with mushroom hunters.

Identification   Cap 3–12″ wide; dry; white, covered with a dense layer
of nonremovable reddish-brown scales. Gills white,
closely set, and free. Stalk 5–8″, sometimes as tall as
16″, with movable ring near top; covered with brown
scales and has bulbous base.

Edibility   Choice, with caution; resembles toxic species.

Similar Species   Toxic Green-spored Lepiota *(Chlorophyllum
molybdites)* develops greenish gills. Toxic
Leucocoprinus brunnea turns slowly brown if sliced;
sliced Parasol turns pink or reddish orange.

Habitat   Under hardwoods or conifers in woods; on lawns.

Range   Eastern North America into the Midwest.

Lilac Geophylla *Inocybe geophylla* var. *lilacina* ⊗

The genus *Inocybe* contains dozens upon dozens of small species, and many of them share visible characteristics to such an extent they may be identified with certainty only by microscopic examination. A number of them, including the Lilac Geophylla, are known to be poisonous. Despite its variability in color, this species is readily recognized in the field. Also known as *Inocybe lilacina*.

Identification	Cap ⅜–1½″ wide; conic, bell-shaped, or expanded with a central bump; dry but slightly sticky in moist weather. Dark lilac or pale lilac centrally, paler elsewhere, fading to faint lilac-gray. Gills closely set, whitish to grayish brown. Narrow stalk to 2¼″, tinged lilac or paler, with a thin ring.
Edibility	Poisonous.
Similar Species	Distinctive when young; faded specimens of the Lilac Geophylla closely resemble other species of *Inocybe*.
Habitat	Under hardwoods or conifers. Sometimes on lawns.
Range	Widely distributed across North America.

38

Mica Cap *Coprinus micaceus*

This small species can be surprisingly difficult to identify. The small, shiny particles on the cap, which give it its common name, are seen only on young mushrooms. Easily washed off by rain, they are almost never found on mature specimens. In addition, although the Mica Cap grows on wood, it may grow from hidden, buried wood, giving it the appearance of a terrestrial mushroom. Some edible members of this genus cause gastric distress if taken with alcohol.

Identification	Cap 1–2¼″ wide; cylindrical to usually bell-shaped, with deep striations (lines) from cap edge inward. Yellow to light brownish, darker centrally, and initially coated with shiny, white granules. Gills closely set; white at first, then purple, then oozy black. Stalk to 3″ long; white; slender and cylindrical.
Edibility	Edible.
Similar Species	Gills of *C. disseminatus* do not form black ooze.
Habitat	Dense clumps around stumps or on roots.
Range	Widely distributed in North America.

Bleeding Mycena *Mycena haematopus*

Although this mushroom is not strikingly colored, it may be the best known of all *Mycena* species. It is common in both the spring and the fall, and is easily identified because a blood-red liquid oozes from the stalk when it is broken or cut.

Identification
Cap ½–2″ wide, with a scalloped edge. Bell shaped or conical, widening with age. Reddish brown and darker centrally with a paler, striate (lined) margin. Usually moist. Closely or widely set gills whitish to pinkish gray with reddish-brown stains. Stalk to 4″; narrow, reddish brown, oozing blood-red juice if broken or cut; hairy at base.

Edibility
Edible, but poorly rated.

Similar Species
No other *Mycena* species in same habitat has stalk liquid. The Brown Dunce Cap *(Conocybe tenera)* lacks stalk liquid, grows on ground.

Habitat
Usually clustered on decaying wood of deciduous trees.

Range
Widely distributed throughout North America.

Golden Trumpets *Xeromphalina campanella*

The common name suggests all the brassy fanfare of grand opera, but one would not get much of a blast out of these tiny mushrooms. What they lack in size, however, they make up for by the sheer number of individuals appearing in one cluster. They can make a surprisingly bright display in the normally grayish-brown environment in which they are found. Also called Fuzzy Foot.

Identification	Cap ¼–1″ wide; yellow-brown to orange-brown; moist and centrally depressed. Gills yellow or dull orange; fairly well separated, running partway down stalk. Stalk ⅜–2″ tall; thin, tough, yellow above, reddish brown below, with a tuft of yellow-brown hairs at base.
Edibility	Unknown; too small and tough to be interesting.
Similar Species	*X. kauffmanii* found on hardwood logs and stumps. Less common species are best separated microscopically.
Habitat	Stumps and logs of conifers, usually well-decayed wood.
Range	Widely distributed in North America.

44

Orange Mycena *Mycena leaiana*

With a little field experience, it is often possible to determine that a mushroom is some type of *Mycena*. Further identification frequently requires microscopic examination. The brilliant Orange Mycena is an exception, as it is distinguished by its habit of forming dense clusters on decaying logs.

Identification	Cap ⅜–2″ wide; bright orange to reddish orange, fading with age. Bell shaped to centrally depressed, and a bit sticky. Gills closely set, with darker reddish-orange edges. Stalk 1¼–2¾″; narrow; orange to yellow with orange fibers at base.
Edibility	Unknown.
Similar Species	The Velvet Foot *(Flammulina velutipes)* has blackish-brown lower stalk.
Habitat	Clustered on decaying hardwood logs, especially beech.
Range	Eastern North America into the Midwest.

Yellow Waxy Cap *Hygrocybe flavescens*

Like the Parrot Mushroom, this beautiful species—also known as the Golden Waxy Cap and *Hygrophorus flavescens*—undergoes color changes as it matures. The display, however, is much less grand. The Yellow Waxy Cap sometimes occurs in great abundance; although it is reportedly edible, most mushroom hunters consider it too small to be interesting.

Identification   Cap 1–2¾″ wide; faintly lined and sticky. Orange at first, then orange-yellow to yellow or paler; gills yellowish. Stalk to 3″ tall; mostly yellow, with some orange tinting, but with a white base. Dry to moist.

Edibility   Edible.

Similar Species   *Hygrocybe chlorophana (Hygrophorus chlorophanus)* has sticky cap and stalk, lacks white at stalk base. *Hygrocybe nitida (Hygrophorus nitidus)* has centrally depressed cap.

Habitat   On the ground under hardwoods, conifers, or in mixed woods, frequently in mossy areas.

Range   Widespread across North America.

Jack O'Lantern *Omphalotus illudens* ⊗

The toxic Jack O'Lantern has bioluminescent gills; a fresh specimen will give off an eerie glow in a darkened room. The cause of this bioluminescence has been the subject of many arguments, but that odd feature may be no more than a by-product of the plant's metabolism. The Jack O'Lantern is called *O. olearius* by some.

Identification   Cap 3–8″ wide; orange to yellow-orange, smooth, and dry; with a small, central bump and usually a sunken center. Closely set orange-yellow gills run down stalk. Stalk to 8″, ringless; light orange; tapers to a narrow, normally fused, base. Unpleasant or sweet odor.

Edibility   Poisonous.

Similar Species   The Chanterelle *(Cantharellus cibarius)* grows singly or in groups (not clusters) on soil. Big Laughing Gym *(Gymnopilus spectabilis)* is ringed. Other *Gymnopilus* species have rings or darker caps, or grow on soil.

Habitat   In clusters on hardwood stumps, trunks, or roots.

Range   Throughout the East and South; also California.

Sulfur Tuft *Naematoloma fasciculare* ⊗

This species is a favorite of more advanced mushroom hunters because it represents some major difficulties in mushroom identification. It resembles several other species, and although in North America it is known to be poisonous, some individuals eat it anyway.

Identification   Cap 1–3¼" wide; smoothly rounded, sometimes with a prominent low bump, or almost flat. Orange-yellow or yellow with olive tones with age. Moist and smooth. Gills closely set, sulfur-yellow to greenish yellow when young, tinted purple-brown in age. Stalk to 4½" long; slender; orangish above, rusty below, with thin, hairy ring.

Edibility   Poisonous.

Similar Species   Cap of Brick Tops *(N. sublateritium)* is brick-red. Gills are grayish to purple-brown in Smoky-gilled Naematoloma *(N. capnoides)*. Deadly Galerina *(Galerina autumnalis)* distinguished by spore print color.

Habitat   Clustered on logs and stumps.

Range   Widely distributed throughout North America.

Yellow Unicorn *Entoloma murraii*

Like wildflowers, fungi display a wide array of pigments; certain experts would contend that some mushrooms surpass wildflowers in beauty. The brilliantly colored Yellow Unicorn, which brightens the damp, lowland woods, is one mushroom that inspires great appreciation among connoisseurs.

Identification	Cap ½–1½″ wide; bright yellow; bell-shaped to conical with a short, pointed peg at center (as if a narrow stalk pierced the top of the cap). Well-spaced gills are yellow first, pink later. The narrow stalk, up to 4″ long, is dry, yellow, and ringless.
Edibility	Unknown.
Similar Species	Salmon Unicorn Entoloma *(E. salmoneum)* structurally similar but has salmon-orange cap and orange stalk.
Habitat	Damp soil in deciduous or mixed woods; also in swamps.
Range	Widely distributed south and east of the Great Lakes.

Liberty Cap *Psilocybe semilanceata* ⊗

The Liberty Cap is one of a small group of mushrooms that induce altered states of consciousness when ingested. Some persons collect them for this reason. Psychotropic species are mostly LBMs ("Little Brown Mushrooms") with some dangerously toxic twins, including the Deadly Galerina, and too small to be of interest to most mushroom hunters.

Identification Cap, typically unexpanded, 1″ high and ¾″ wide; conical, sometimes bell-shaped, frequently with a small, knobby tip. Sticky, smooth, and striate (lined) when moist. Dark chestnut brown, fading to yellow-brown or paler, tinged olive or green. Gills closely set, mature purple-brown. Stalk to 4″ long; slender; pale above, is reddish brown below, staining blue or greenish at base.

Edibility Poisonous.

Similar Species Conifer Psilocybe *(P. pelliculosa)* and *P. silvatica* are found primarily in conifer areas.

Habitat In grassy farmyard fields and well-manured meadows.

Range Common in the Pacific Northwest; rare elsewhere.

Violet Cortinarius *Cortinarius violaceus*

Authorities estimate there may be as many as 1,000 species of *Cortinarius* in North America. Out of this vast assemblage, hundreds of species are known to possess varying degrees of violet or lilac pigments. Even so, the Violet Cort is consistently singled out for the vividness of its color.

Identification	Cap 2–4½″ wide; dry; sometimes with a low bump at center. Deep violet, sometimes shiny; covered with small, scaly tufts of hair. Gills widely spaced, dark violet, becoming covered with rusty powder with age. Stalk 2¾–6″ or taller; stout, dry, fibrous; deep violet, often with a thin, webby ring.
Edibility	Edible with caution; misidentifications are frequent.
Similar Species	Distinguished from other corts by deep violet tones, and dry, rough cap. Viscid Violet Cort *(C. iodes)* has sticky cap.
Habitat	On the ground under conifers.
Range	Northern half of United States into Canada.

Old Man of the Woods *Strobilomyces floccopus*

The Old Man of the Woods is also commonly called the Pine Cone Fungus, a direct translation of the genus name *Strobilomyces*. Those persons accustomed to finding typical boletes may be surprised to discover this shaggy member of the clan. The species pictured is very common; others in the genus resemble this one so closely they are best separated microscopically.

Identification Cap 1½–6″ wide; heavily scaled and shaggy; grayish black, mixed with paler gray. Lower surface of cap pored, usually ash-white. Stalk to 4″ tall, scaly and hairy, dark gray, with a shaggy ring.

Edibility Edible, but uninteresting as table fare.

Similar Species Two relatives, *S. confusus* and *S. dryophilus* (predominantly in the Southeast) are very similar.

Habitat Under hardwoods or in mixed woods, or on rotting wood.

Range Eastern North America into the Midwest.

Poison Pie *Hebeloma crustuliniforme* ⊗

When this species occurs in an open area, it may form a circle of mushrooms, called a fairy ring, in the fashion of *Marasmius oreades*. This particular pattern of occurrence must not be taken to indicate edibility; for the *Marasmius* is edible but the *Hebeloma*—as its wryly explicit common name declares—is not.

Identification
Cap 1–3½″ wide; sticky, with a low, central bump. Brown or reddish brown at center, creamy buff elsewhere. Closely set gills whitish when young, dull brown later, sometimes beaded with moisture over finely toothed edges. Stalk, to 3″, whitish with fine particles above; sometimes widening toward base. Smells like radishes.

Edibility
Poisonous.

Similar Species
Gills of *Tricholoma* species may stain brown but are typically paler. Young *Cortinarius* gills cobwebby.

Habitat
In open woods, under conifers and hardwoods; in fields.

Range
Widely distributed in North America.

Poison Powderpuff *Lactarius torminosus* ⊗

All members of the genus *Lactarius* emit a fluid (called "milk") when the gills are sliced. It is important to note the color as the flow begins, and whether or not the color changes. The Poison Powderpuff has milk that starts out white and stays that way. While this mushroom is considered poisonous in North America, it is eaten in some places in Europe. This contradiction may indicate that different strains of the species are involved. Also known as Pink-fringed Milky.

Identification	Cap to 4½″; pinkish and depressed centrally, with a paler, hairy-fringed margin. Gills close together; yellowish pink with white, unchanging milk. Stalk to 2¾″, same color as cap, occasionally spotted.
Edibility	Poisonous.
Similar Species	Older and faded examples of this species may be mistaken for a number of other mushrooms whose edibility is not yet determined.
Habitat	On the ground, usually associated with birch.
Range	Widespread, primarily in the North; also in California.

Meadow Mushroom *Agaricus campestris*

This common edible mushroom (also commonly called Pink Bottom) is one of the most frequently collected in North America. Nonetheless, beginners especially should beware toxic look-alikes. Like *Agaricus* species, *Amanita*, *Lepiota*, and *Pluteus* are free-gilled; the gills may be whitish, yellow, pink, or red, but not brown.

Identification

Cap 1–4″ wide; flat or gently convex, and dry; silky-smooth or somewhat scaly; white to pale gray-brown. Gills closely set, free; pink, becoming purple-brown. Stalk to 2″; stout; white with brown stains, thin ring.

Edibility

Choice.

Similar Species

Toxic California Agaricus *(A. californicus)* has gills whitish at first. Toxic Felt-ringed Agaricus *(A. hondensis)* has a well-formed, flaring ring. Toxic Yellow-foot Agaricus *(A. xanthodermus)* bruises yellow, smells bad. Deadly *Amanita* spp., often in same habitat, have white spore print.

Habitat

Grassy areas, meadows, barnyards, and along roads.

Range

Widely distributed throughout North America.

White Matsutake *Armillaria ponderosa*

The exotic common name of the White Matsutake indicates its relationship to an Oriental fungus. It is a very popular edible on the West Coast, where it is not only collected on an informal basis but for the commercial market as well. It is also known under the name *Tricholoma magnivelare*.

Identification   Cap from 2–8″ broad; white when young, developing flat, reddish-brown scales centrally, with streaks of brownish fibers elsewhere. Moist or somewhat sticky. Gills closely set, white turning reddish brown when rubbed. Stalk stout, 2–6″ tall, with a prominent ring; white above the ring, with brown hairs and scales below it. Fragrant odor.

Edibility   Choice.

Similar Species   Swollen-stalked Cat *(Catathelasma ventricosa)* has an ash-gray cap. *Hygrophorus subalpinus* has gills that run partway down the stalk.

Habitat   Under conifers, or in thickets in coastal sandy soil.

Range   Entire northern area; abundant on U.S. West Coast.

Anise Clitocybe *Clitocybe odora*

The common name of this beautiful and variable-hued mushroom refers to its aniselike odor, which is almost always present and can be quite strong. This well-known edible is seldom eaten alone but is usually mixed with other, less flavorful mushrooms.

Identification	Cap 1–4″ wide; usually bluish green, but may be dingy green, pale bluish gray, solid blue, or whitish; smooth and moist, and flat to somewhat depressed. Closely set gills are whitish, buff, or green-tinged (bluish-green in the western variety *pacifica*), and may run slightly down the stalk. Stalk to 3½″ tall, cylindrical; white, buff, or greenish; usually wider at base. Smells like anise.
Edibility	Edible.
Similar Species	Blue gills of the Indigo Milky *(Lactarius indigo)* reveal blue fluid if cut.
Habitat	On the ground under hardwoods or in mixed woods.
Range	Widely distributed in North America.

Blewit *Clitocybe nuda*

The odd common name, which means "blue hat," indicates the Blewit's chief clue to identification. When the cap fades, its violet tones give way to a buff-brown color. The gills and stalk, however, almost always retain an infusion of violet, insuring that the mushroom at any stage is clothed in vivid color.

Identification   Cap 2–6″ broad, violet when young, fading to buff-brown or tan, smooth and moist or dry. Closely set violet or buff gills may run partway down stalk. Stalk, to 4″, pale violet usually flecked gray or brown; has bulbous base. Often has fragrant odor.

Edibility   Choice, with caution.

Similar Species   *Cortinarius* species have a webby veil over young gills. Other *Clitocybe* species are very close and may be merely varieties. A few *Entoloma* spp. similar.

Habitat   Under hardwoods or conifers, in grass, or on compost heaps.

Range   Widely distributed across North America.

Saffron Parasol *Cystoderma amianthinum*

Sometimes a mushroom with the most pleasant appearance, such as the Saffron Parasol, can stir up a controversy behind the scenes. This attractive species was at one time considered to be a *Lepiota*, and it was then later classified in *Armillaria*. Now classified as a *Cystoderma*, it quietly brightens the woodland floor in late summer and fall. Also called Pungent Cystoderma.

Identification
Cap 1–2″ wide; yellow-brown to yellow-orange, often with a central bump, and coated with granules. Flaps of tissue at cap edge. Buffy gills are closely set. Stalk to 3″, with buff area above a poorly formed ring; similar to cap below ring. May smell like corn.

Edibility
Unknown, and too small to be interesting.

Similar Species
Darker Common Conifer Cystoderma *(C. fallax)* has well-developed ring. *C. granosum* grows on decaying hardwoods. *C. cinnabarinum* is larger with pinkish tones in cap.

Habitat
In mossy needle beds or on humus in conifer woods.

Range
Widely distributed. Common in the Pacific Northwest.

Aborted Entoloma *Entoloma abortivum*

This species piles peculiarity upon peculiarity. It occurs in both a normal and an abortive form; and although both are technically edible, neither is recommended, as each resembles toxic species. When both types are found growing together, however, it is easier to be sure of their identity. The abortive type results from the presence of another fungus, *Armillaria mellea*.

Identification
: Cap 2–4″ wide, flat or with a central bump, gray or gray-brown and dry. Closely set gills gray at first, later pink, may run down stalk a bit. Solid stalk to 4″ long, white to grayish. Abortive form has irregular, lumpy white cap, 1″ or larger, somewhat marbled internally.

Edibility
: Edible, but not recommended.

Similar Species
: Normal form alone resembles toxic *Entoloma* species, only distinguishable microscopically. Abortive form resembles immatures of toxic *Amanita* species. When both grow together, they are unmistakable.

Habitat
: On soil, and on or near well-rotted logs and stumps.

Range
: Throughout eastern North America and west to Texas.

Blusher *Amanita rubescens*

"Rubescens" means "becoming red," which explains the common name of this mushroom: the Blusher develops red stains when rubbed or sliced. In fact, it characteristically shows such markings while it is still in the ground. The Blusher is edible; but in keeping with a better-safe-than-sorry philosophy, it cannot be recommended as table fare.

Identification   Cap 2–6″ wide; centrally knobbed, reddish brown. Usually a bit sticky, with scattered, removable, olive-gray, white, or pinkish patches. Gills closely set, free; white with reddish stains. Pinkish stalk, 3–8″, has skirtlike ring above, with red stains at the usually bulbous base. Torn bands and patches, remnants of basal cup, usually present.

Edibility   Edible, but not recommended.

Similar Species   Toxic Panther *(A. pantherina)* has more persistent, rimmed basal bulb.

Habitat   Found under hardwoods or white pine.

Range   Eastern North America; also in California.

Poison Paxillus *Paxillus involutus* ⊗

It is generally accepted that mushrooms should never be eaten raw. Some individuals, however, have eaten raw Poison Paxillus mushrooms for years with no ill effect. This mushroom has caused fatalities in Europe, and it is now believed to cause kidney failure due to a gradually acquired hypersensitivity.

Identification	Cap 1½–5″ or wider; almost flat to somewhat depressed at center; furrowed and riblike near edge, with rim turned under. Yellow-brown to dark brown with flat hairs, and dry but sticky when moist. Gills very closely set; olive-yellow, run partway down stalk; stain brown if rubbed, and separate easily from cap. Stout stalk, to 4″ long, yellow-brown to dark brown, and dry.
Edibility	Poisonous.
Similar Species	Velvet-footed Pax *(P. atrotomentosus)* has velvety dark brown stalk. *Lactarius* species have gills that exude milk if sliced.
Habitat	Under conifers or in mixed woods; in lawns; on wood.
Range	Widely distributed in North America.

Scaly Hydnum *Sarcodon imbricatum*

The common name of the Scaly Hydnum harks back to the time when mushrooms with spines on the underside of the cap were all classified in the genus *Hydnum*. Subsequent research has created many new genera, including *Sarcodon*, into which this species has been placed. This large, common mushroom has a flavor that varies from mild to unpleasant; the species is known to have made a small number of individuals ill.

Identification	Cap up to 8″ wide; covered with coarse, brown scales; usually depressed at center. Underside covered with crowded brown spines. Stalk pale brown, up to 4″ long, and smooth. Mushroom dry overall.
Edibility	Edible; but has caused illness in some individuals.
Similar Species	*S. scabrosum* has smaller scales; is usually tinted red and tastes bitter.
Habitat	On the ground in both hardwood and conifer forests.
Range	Widely distributed across North America.

Wine-cap Stropharia *Stropharia rugoso-annulata*

The species name of this mushroom (which is sometimes spelled without a hyphen) means "wrinkled ring." It refers to the lower portion of the ring, which is split into many outwardly directed and upturned segments, like tiny hooks. This species thrives on wood chips, and produces a bumper crop when it occurs.

Identification: Cap 2–6″ wide, in a long, low curve or flat. Normally smooth or slightly roughened. Purplish brown or reddish brown, fading to tan or grayish yellow, and dry. Gills closely set, white at first, then gray-blue, and finally deep purplish brown. Stalk to 6″ long; stout; white or pale gray; well-formed split ring has upcurled segments on its lower edge.

Edibility: Choice.

Similar Species: Similar *Agrocybe* spp. have brown gills; *Agaricus* spp. have free gills.

Habitat: On wood chips around plantings and in woods.

Range: Widespread across northern North America.

Purple-gilled Laccaria *Laccaria ochropurpurea*

The Purple-gilled Laccaria has been described as a difficult mushroom to identify, but when it has been seen two or three times, it can easily be distinguished by its robustness, pale color, and, when overturned, deep purple gills. It will occasionally fruit abundantly.

Identification Cap 2–5″ wide; dry and smooth to slightly rough. Initially purplish, fading rapidly to ocher or nearly white. Widely spaced gills are deep purple, fading paler. Stalk up to 6″ long; dry, solid, ringless, sometimes curved; colored like cap, perhaps with brown stains.

Edibility Edible, but not highly regarded.

Similar Species Young *Cortinarius* specimens have webby veil over gills and usually mature to rusty, reddish, or violet colors.

Habitat On the ground in thin, hardwood forests and in grassy areas generally.

Range North America east of the Rocky Mountains.

Bracelet Cortinarius *Cortinarius armillatus*

When immature, all species of *Cortinarius* have a thin, webby layer of fibers extending from the top of the stalk to the edge of the cap, covering the gills. As they grow, the expanding cap tears the web, sometimes leaving a thin ring on the stalk that may disappear later. The distinctive markings on the stalk of the Bracelet Cort form as the stalk expands and an outer layer is stretched, ending as a series of bands.

Identification	Cap 2–5″ broad, usually with a long, low bump; reddish brown to reddish orange; moist; smooth or covered with tiny hairs. Gills well spaced; light brown at first, becoming reddish orange to rusty later. Stalk to 6″ long; stout with bulbous base; dull brownish; encircled by one to several reddish bands.
Edibility	Edible with caution; accurate identification is critical.
Similar Species	Deadly toxic Poznan Cort *(C. orellanus)* and Deadly Cort *(C. gentilis)* have more cylindrical stalk.
Habitat	On the ground in mixed forests, especially near birch.
Range	Northern North America; uncommon westward.

Bluing Bolete *Gyroporus cyanescens*

This somewhat unprepossessing mushroom can provide a spectacular color show for those who discover it. If you pick this mushroom and rub its surface, it will stain blue. However, if the mushroom is sliced quickly from top to bottom in longitudinal sections, it is transformed instantaneously from nearly white to deep indigo blue. The stalk is hollow.

Identification   Cap to 4½″ wide, slightly roughened or velvety, straw colored to almost white. Undersurface pored, white to yellowish. Stalk, to 4″ tall, usually the same color as cap. All parts instantly turn blue if rubbed or sliced.

Edibility   Edible, but not recommended.

Similar Species   The Bluing Bolete is usually distinctive, but several other similar boletes, including toxic Red-mouthed Bolete *(Boletus subvelutipes)* also turns blue instantly if rubbed or sliced.

Habitat   Usually in sandy soil under hardwoods, or in mixed woods.

Range   Eastern North America.

90

Man on Horseback *Tricholoma flavovirens*

With cooperative weather, this fine edible species extends the mushroom-hunting season into December. Like the Sandy Laccaria, Man on Horseback is sticky and normally grows in sandy areas under pine. Also called the Canary Trich and *T. equestre.*

Identification
: Cap 2–4½″ broad, slightly sticky, yellow with reddish-brown center increasing in size with age; usually with a long, low bump. The closely set gills are solidly yellow. Stout, solid stalk (but hollow when old) may be 3½″ long, is sometimes widened below, and is pale yellow to almost white. Odor is not disagreeable.

Edibility
: Edible with caution; resembles toxic species.

Similar Species
: Toxic Sulfur Trich *(T. sulphureum)* is sulfur-yellow, not sticky, with strong, unpleasant odor. Toxic Separating Trich *(T. sejunctum)* is greenish yellow with dark fibers on cap.

Habitat
: Usually under pines in sand or moss, and under hardwoods.

Range
: Widely distributed across North America.

Almond-scented Russula *Russula laurocerasi*

Despite its name, the Almond-scented Russula has an extremely disagreeable flavor. Its odor does, indeed, suggest almonds or maraschino cherries, but it has an unpleasant, fetid component. This species is robust and occasionally abundant.

Identification Cap from 1–5″ wide, yellow-brown or yellow-tan; usually moist and shiny; striate (lined) at the rim. Gills closely set to widely spaced; off-white or pale orange. Stalk about 4″ tall; cylindrical, yellowish white.

Edibility Inedible.

Similar Species *R. subfoetens* smaller, darker, with more fetid odor; *R. fragrantissima* strongly fetid, darker; both best distinguished microscopically.

Habitat On the ground in mixed woods.

Range Widely distributed in eastern North America.

Sandy Laccaria *Laccaria trullisata*

To those persons interested in collecting and eating wild mushrooms, the Sandy Laccaria poses an interesting problem. This robust species thrives in almost barren locales, such as sand dunes. As a result, sand grains are as much a part of the mushroom as the fungal tissue itself, rendering an otherwise edible mushroom too sandy to eat.

Identification — Cap 1–3″ wide; reddish brown, fading to grayish pink; slightly roughened, moist and slightly sticky. Gills well-spaced; purple, purple-pink, or pink, sometimes running slightly down stalk. Stalk to 4″; stout, somewhat enlarged below; colored like the cap. Sand on all parts.

Edibility — Edible, but useless.

Similar Species — None. Distinguished by sandy habitat.

Habitat — Sand dunes, sandy streamsides, and other very sandy areas.

Range — Great Lakes to Atlantic and Gulf coasts.

Orange-capped Bolete *Leccinum aurantiacum*

This mushroom, once placed in the genus *Boletus*, is a complex of closely related species that have not been systematically studied. No *Leccinum* species is known to be toxic, but not all have been tested, so use caution. Also called the Red-capped Scaber Stalk.

Identification
Cap 2–8″ wide; reddish orange or rusty red. Undersurface is pale yellow-brown, marked with pores; turns olive when rubbed. Cap has white flesh, turning wine-red to dark purple-gray when sliced. Stalk 4–8″ long, cylindric or somewhat swollen; and with rough, rigid brown projections that turn black with age.

Edibility
Good.

Similar Species
White flesh in sliced cap of Aspen Scaber Stalk (*L. insigne*) turns dark smoky brown or grayish violet, not wine-red. Common Scaber Stalk (*L. scabrum*) has a gray-brown to yellow-brown cap.

Habitat
On the ground, usually under aspen and pines.

Range
Northern North America, California, and Colorado.

King Bolete *Boletus edulis*

One may judge from its common name the high esteem in which this choice, edible mushroom, a well-known international favorite, is held. This species undergoes subtle geographic variation in stature and color, but all the forms are good to eat. It is one of a group of mushrooms with pores on the underside of the cap; collectively they are known as boletes.

Identification	Cap from 3–10″ wide; bun shaped. Medium reddish brown, varying paler or darker. Lower surface of cap white first, greenish yellow later, and covered with pores. Stalk bulbous, 4–8″ tall, with a fine, white network, sometimes indistinct, on the upper part.
Edibility	Choice.
Similar Species	The Bitter Bolete *(Tylopilus felleus)* has pinkish pore surface. Many other boletes are generally similar.
Habitat	On the ground in coniferous and deciduous woods.
Range	Widely distributed throughout North America.

100

Gray Amanitopsis *Amanita ceciliae*

As is reflected in its common name, this species was formerly placed in the genus *Amanitopsis*, a group of mushrooms similar to the *Amanita*s in most ways, but lacking a ring on the stalk. Members of the two genera, however, share so many other characteristics that all are now placed in the genus *Amanita*. This species is also called the Strangulated Amanita *(A. inaurata)*.

Identification
Cap 2–5″ wide; grayish brown to gray, grooved at the rim, with scattered, removable gray patches. Gills white, closely set, free. Stalk to 6″ tall; grayish white, without a ring; frequently with faint chevron patterning; grows from basal cup, but cup may disintegrate.

Edibility
Reportedly edible but not recommended.

Similar Species
Poisonous Panther *(A. pantherina)* similar; ring on stalk may disappear. Best separated from other ringless *Amanita*s by microscopic features.

Habitat
Usually under conifers; also in mixed open woods.

Range
Widely distributed throughout North America.

The Prince *Agaricus augustus*

Few mushrooms have earned a more impressive name than the Prince, which may be attributed to its noble stature and its reputation as a choice edible.

Identification
Cap 5–10″ wide, sometimes to 15″; often depressed at center; dry. Covered with small, yellow-brown or pale reddish-brown scales, staining yellow when rubbed or sliced. Gills closely set, free; stalk white at first, then pink, finally chocolate-brown. Stalk to 8″ long; stout, with well-formed ring above and delicate, red-brown scales below. May smell of almond or anise.

Edibility
Choice.

Similar Species
Toxic Felt-ringed Agaricus *(A. hondensis)* lacks scales below ring. Toxic *A. praeclaresquamosus* has darker scales. Edible *A. subrufescens* and *A. perrarus* must be separated microscopically for certain identification.

Habitat
Along roadsides, in grassy areas, and in conifer woods.

Range
Rocky Mountains and West Coast. Less common in the East.

Scaly Pholiota *Pholiota squarrosa* ⊗

Edibility in fungi is always an uncertain business. Options regarding the Scaly Pholiota range from "one of the best" to "unpalatable." Some people report that the species is mildly poisonous. Clearly it is best avoided.

Identification Cap from 1–4″ broad, with a central bump or flat surface; yellow, densely covered with yellowish-brown scales; dry. Closely set gills greenish-yellow when young, becoming dingy brown. Stalk to 4″ long; scaly, with yellow ring; colored like cap; dry.

Edibility Poisonous.

Similar Species *Pholiota squarrosa-adiposa* has a sticky cap surface. Yellow Pholiota *(P. flammans)* yellow overall with yellow scales, and sticky. Sharp-scaly Pholiota *(P. squarrosoides)* has sticky surface; and between scales whitish, not yellow.

Habitat Clustered at bases of trees; also on logs and stumps.

Range Widely distributed. Common in the Rocky Mountains.

Rosy Gomphidius *Gomphidius subroseus*

Looking like giant nails driven into the earth, the Rosy Gomphidius is a welcome sight on two counts. Not only is this an edible species, but its appearance—at least in the western part of its range—signals the beginning of the main mushroom collecting season. Some persons find its flavor mediocre, while others consider it choice.

Identification   Cap 2–4″ across, and pink, rose-pink, or red. Sticky to slimy. Close to well-spaced gills are initially white but pale smoky-gray later. Gills run down the stalk, and some are forked. The stout stalk may exceed 3″, is white above and yellow below, and it usually has a pale, sticky ring that darkens later.

Edibility   Edible to choice.

Similar Species   Other *Gomphidius* species are colored differently.

Habitat   Under conifers, particularly Douglas-fir.

Range   Widely distributed in northern North America; south in mountains to California and North Carolina.

Indigo Lactarius *Lactarius indigo*

The Indigo Lactarius is one mushroom that is easy to identify. This edible species has a blue cap, blue gills, and blue stalk. Even the milk from the sliced gills is blue. It is abundant only in the southeastern portion of its range, and it is a bit coarse in texture.

Identification	Cap 2–6″ wide, centrally depressed; Dark blue, turns paler with age. Gills dark blue, aging paler, releasing blue milk when sliced. Stalk to 3¼″, dark blue or silver-blue.
Edibility	Edible; reportedly coarse but good.
Similar Species	Silver-blue Milky *(L. paradoxus)* has pale orange to purplish-red gills and brownish-purple milk.
Habitat	On the ground in hardwood or coniferous forests.
Range	Widely distributed from Texas and the Midwest to the East Coast. Most abundant in the Southeast.

Delicious Lactarius *Lactarius deliciosus*

Some edible mushrooms seem to have built-in safeguards that discourage their being eaten by mushroom hunters. The Delicious Lactarius, which is orange, develops suspicious green stains over its surface as it grows; this unsightly camouflage has thwarted more than one collector from picking it. Also called the Orange-latex Milky.

Identification   Cap up to 5″ broad; carrot-orange, usually with green stains. Centrally depressed. Gills closely set, orange, staining green. Sliced gills reveal carrot-orange milk. Stalk to 2½″, light orange with some green.

Edibility   Edible; considered choice by some.

Similar Species   *L. thyinos*, found in northern conifer bogs, has sticky stalk and does not stain green. *L. sanguifluus* has blood-red milk. Peck's Milky *(Lactarius peckii)* has white milk.

Habitat   Under coniferous trees, especially pine.

Range   Widely distributed throughout North America.

Gilled Bolete *Phylloporus rhodoxanthus*

The common name for this species is confusing, because it seems contradictory—everyone knows that boletes have pores, not gills, on the cap undersurface. In spite of this divergence, the species shares some bolete characteristics. The lower layer of the cap may be easily separated from the upper layer (as is true of nearly all boletes). When you make this experiment, you will occasionally find pores among the gills.

Identification   Cap to 3″; yellow-brown or red-brown, and dry. Lower surface of cap with well-spaced, bright yellow gills running down stalk; gills may be partly pored. Stalk to 3″ tall, buffy with reddish tints.

Edibility   Edible but not recommended; resembles toxic species.

Similar Species   Toxic Poison Paxillus *(Paxillus involutus)* has similar pored gills descending stalk, but they are closer together.

Habitat   On the ground in mixed woods or under conifers.

Range   Widely distributed in North America.

Tawny Milkcap *Lactarius volemus*

The Tawny Milkcap frequently develops a rather unpleasant fishy odor shortly after being picked. Moreover, the whitish gills stain an unsightly brown from the freely flowing milk that is released when the gills are sliced. Despite these features, the Tawny Milkcap is an edible species that is popular with mushroom hunters.

Identification   Cap 2–5"; orange-brown, centrally depressed and smooth. Gills close together, almost white. Copious white milk flows from sliced gills, staining them brown. Stalk up to 4" tall, buffy orange with darker stains.

Edibility   Choice.

Similar Species   Corrugated-cap Milky *(L. corrugis)* has a wrinkled cap; Hygrophorus Milky *(L. hygrophoroides)* has widely spaced gills and does not develop a fishy odor or stain brown. Both are good edibles.

Habitat   On the ground in deciduous woods, or in mixed woods.

Range   Eastern North America west to Texas.

Red Chanterelle *Cantharellus cinnabarinus*

As colorful as any wildflower, the Red Chanterelle will occasionally form a reddish-orange carpet in the woods in late summer. Its overall shape, which suggests a small funnel or trumpet, and the blunt, gill-like ridges on the underside, are excellent field marks of the genus *Cantharellus*. When you add the brilliant color, the identification to species is complete.

Identification   A small, shallow funnel to 2½″ tall. Cap ½–1½″ across, reddish orange or cinnabar-red, fading in sunlight to pinkish white. Blunt, gill-like ridges run down the stalk in an open, pinkish network. Stalk colored like cap, 1½″ tall, ¼″ wide.

Edibility   Edible, but not of high quality.

Similar Species   Red *Hygrophorus* species have well-formed gills with sharp edges; gills do not form an open network.

Habitat   On the ground in open woods, especially oak, and along woodland paths.

Range   Eastern North America.

Larch Waxy Cap *Hygrophorus speciosus*

Mushrooms in the closely allied genera *Hygrophorus* and *Hygrocybe* are beautiful, but they are notorious among mycologists. Many of these species—including the Larch Waxy Cap—display a wide variety of colors, making identification extremely difficult.

Identification
Cap 1–2″ wide; usually with a central bump, but sometimes with sunken center. Red or orange-red, fading through orange to yellow; covered with a sticky substance. Well-spaced white to yellowish gills, usually with darker edges, run down stalk. Stalk to 4″; white or yellow; sticky at first, but the sticky substance may dry and leave orange stains.

Edibility
Edibility uncertain; best left alone.

Similar Species
Many other *Hygrophorus* species also change color. Late Fall Waxy Cap *(H. hypothejus)* initially olive-brown but ages yellow, orange, or even red-tinted.

Habitat
Under conifers, frequently larch, in moist areas.

Range
Pacific Northwest, Rocky Mountains, and eastern North America.

Frost's Bolete *Boletus frostii*

This species—named to honor an early American mycologist, Charles Frost—is considered to be one of the most beautiful of all fungi. The attractive reddish color of this species, which fruits in summer and early fall, is at its most intense before the mushroom is fully mature.

Identification	Cap 2–6″ wide; sticky; blood-red or somewhat paler. Undersurface red, with pores, at times with yellow droplets; turns blue if rubbed. Stalk to 4″ or more; blood-red, coarsely webbed, and lacerated. Flesh yellow, turning blue when sliced.
Edibility	Reportedly edible but definitely not recommended. Despite its distinctive stalk, it resembles toxic species.
Similar Species	Poisonous Red-mouth Bolete *(B. subvelutipes)* and close allies also have a reddish pore surface and yellow flesh that turns blue.
Habitat	On the ground, sometimes abundantly, in oak woods.
Range	Throughout eastern North America.

Sickener *Russula emetica* ⊗

Although generally regarded as poisonous (or at least inedible because of its hot, acrid taste) the Sickener is still eaten by some people. To most collectors, however, the bright red cap flashes a warning. *Russula* species are closely related to *Lactarius* species, but *Russula*s do not exude milk when their gills are sliced and are characterized by their extreme brittleness. This species is also called the Emetic Russula.

Identification	Cap up to 4½″ wide; sticky; bright red, fading with age, and striate (lined) at the rim. Gills closely set, white, and brittle. Stalk up to 4½″; cylindrical, white.
Edibility	Poisonous.
Similar Species	Most red-capped *Russula* are very similar; best distinguished by microscopic examination.
Habitat	In sphagnum moss, under conifers, or in mixed woods.
Range	Widely distributed across North America.

Haymaker's Mushroom *Psathyrella foenisecii* ⊗

Placed in the genus *Panaeolus* by some mushroom experts, this species is well known to those who strive to maintain an uncluttered lawn—and is also known, prosaically, as the Lawn Mower's Mushroom. Primarily a springtime species, it is reported to be both weakly psychotropic and poisonous.

Identification Cap ⅜–1¼″ broad; conical to bell-shaped; reddish brown when moist, drying rapidly to light brown or paler, frequently with a darker band near cap edge. Gills closely set; deep purple-brown. Fragile stalk, to 3″ long or more, whitish or pink-tinged.

Edibility Poisonous.

Similar Species Common Psathyrella *(P. candolleana)* larger, less sturdy, with partial webby veil. *P. castaneifolia* has much thicker stalk and strong odor. *Agrocybe pediades* has yellow-buff cap, paler gills.

Habitat In grassy areas.

Range Widely distributed and common in North America.

Parrot Mushroom *Hygrocybe psittacina*

Also called *Hygrophorus psittacinus*, the Parrot Mushroom displays the colors of tropical feathered finery; like a quick-change artist, it wears them in succession as the mushroom develops to maturity. Some mushroom experts place all members of the Waxy Cap family in the genus *Hygrophorus*.

Identification
Cap expands to just over 1″ wide; usually bell-shaped; deep green at first, fading through orange-yellow to pinkish; sticky, with radial lines. Gills well-spaced, tinted green, red, or yellow. Stalk to 3″ tall; also exhibits same color changes. A blue-capped form is found in California.

Edibility
Edible with caution.

Similar Species
Unmistakable when young; older specimens may be misidentified. *Leptonia (Entoloma) incana* has yellow-green cap, green stalk, and bad odor.

Habitat
On the ground in mixed woods, fields, and along roads.

Range
Widely distributed in North America.

Gem-studded Puffball *Lycoperdon perlatum*

The puffball group includes species ranging in size from those not much larger than a pea to others the size of a bushel basket. All produce spores in an initially sealed enclosure, and most are edible when young and white internally. Not all make good eating, however, and some are purgatives. While no poisonous species are known, only the immature puffballs are edible.

Identification	Mushroom 3″ or more tall, 1–2½″ wide; round, pear-shaped, or top-shaped; covered with long, removable spines, and shorter, permanent spines. Develops center pore with age. Base smooth.
Edibility	Choice.
Similar Species	Distinguished by unique spines. If interior reveals T-shaped pattern when sliced longitudinally, it is a young, perhaps toxic, *Amanita*.
Habitat	Clustered on the ground in mixed woods, and in compost.
Range	Widely distributed throughout North America.

Shaggy Mane *Coprinus comatus*

Unlike most mushrooms, a typical *Coprinus* does not spread its cap widely when releasing spores. Rather, it releases them from the lower part of the downhanging gills and that section then liquefies, forming an inky-black ooze and clearing the way for the next section of the gill to release spores. This life cycle explains why *Coprinus* species are commonly called "Inky Caps."

Identification   Cap 2–4″ tall; cylindrical at first, fluffy-scaly, and white, later with brownish tips to scales as cap expands somewhat. Gills closely set; white when young, purple later, finally turning to black liquid. Narrow, white stalk to 8″, sometimes with fragile ring.

Edibility   Edible. Considered choice by some, disliked by others.

Similar Species   The Scaly Inky Cap *(C. quadrifidus)* clusters on hardwood debris.

Habitat   Grouped in open, grassy areas or hardpacked soil.

Range   Widely distributed in North America.

Collared Earthstar *Geastrum triplex*

The Collared Earthstar is a handsome and striking species, one of several in the genus *Geastrum*. This group—members of the Puffball family—have their spores in a central mass, which is protected by an outer covering. As the mushroom matures, the covering splits into rays, producing a starlike pattern.

Identification
Starlike structure to 3½″ wide, with 4–8 sharply pointed rays; cracked surfaces, surrounding an inflated, saclike structure (spore mass) having a central pore; spore mass surrounded by a bowl-like collar. Color varying from dark brown to pale buff-brown, sometimes almost white.

Edibility
Edible but bland.

Similar Species
Distinguished by its large size and collar. The Rounded Earthstar *(G. saccatum)* is much smaller and lacks a collar.

Habitat
Amid leaves or on rich humus in mixed woods.

Range
Widely distributed in North America.

Bird's Nest Fungus *Crucibulum laeve*

It is widely accepted that mushrooms may assume odd, even bizarre, forms. Nevertheless, nothing seems more improbable than a fungus that resembles a tiny bird's nest, complete with eggs! The Bird's Nest Fungus seen here, and other species with the same form, are members of a group called the Nidulares. They are frequently overlooked due to their small size.

Identification
: Cup-shaped structures, to ⅜″ wide at top and ⅜″ tall, tawny yellow to cinnamon brown outside, covered with velvety hairs. Inner surface pale, smooth, and shiny. Cup contains numerous beadlike pale buff or white objects, each attached to the cup by a tiny cord. Immature cups covered by yellow, hairy membrane.

Edibility
: Inedible.

Similar Species
: Other bird's nest genera (*Cyathus*, *Nidula*, and *Nidularia*) similar but have dark "eggs."

Habitat
: Usually clustered on decaying wood or woody debris.

Range
: Widely distributed throughout North America.

Pigskin Poison Puffball *Scleroderma citrinum* ⊗

Persons who regularly eat wild mushrooms are familiar with the rule that puffballs are edible when white on the inside, but there are two exceptions to this rule. Some thin-skinned puffballs, technically edible, are flavorless or unpalatable. This species, a puffball mimic, often has a white interior, but it is poisonous. Fortunately, it darkens early in its development and has a thick rind, which is distinctive.

Identification	Mushroom to 3″ wide, 2″ tall; more or less round; yellow-brown, with a network of lines and cracks, and raised warts. White inside when very young, becoming marbled, then deep purple-black. Thick white rindlike skin turns pink when cut. Stalk very short or absent.
Edibility	Poisonous.
Similar Species	No true puffball has distinctive embossed pattern of warts and lines on surface.
Habitat	On rich humus in mixed woods, or on very rotten wood.
Range	Widely distributed in North America.

Aspic Puffball *Calostoma cinnabarina*

Many persons shy away from using the Latin name of a mushroom and use instead the more easily memorized common name. "Aspic Puffball" is a fine functional name, but *Calostoma cinnabarina* means "beautiful red mouth," which certainly qualifies as a vivid description. This unusual fungus always evokes comment, especially when first discovered.

Identification	Stalked globe to 2″ tall, initially covered with gelatinous material; this falls away, enveloping stalk in "aspic" containing red, seedlike particles, and revealing red to reddish-yellow globe, to ⅝″ wide, with red central pore. Netted, red, stout stalk to 1¼″ tall.
Edibility	Unknown.
Similar Species	Yellow *C. lutescens* has longer stalk. Ravenel's Stalked Puffball *(C. ravenelii)* and *C. microsporum* lack gelatin.
Habitat	On the ground in woods and along roads.
Range	Eastern United States, west in the south to Texas.

Orange Peel *Aleuria aurantia*

This brightly colored and abundant mushroom normally grows fully exposed. These characteristics, plus the fact that the Orange Peel is palatable—unlike many other cup fungi—are enough to make it popular with some mushroom hunters. Good rains, particularly in the fall, will ensure a crop of large specimens.

Identification Saucer shaped to shallowly cup shaped; up to 4″ in diameter. Orange or orange-yellow on the inner surface, paler on the exterior. Stalkless.

Edibility Edible; reasonably tasty when eaten fresh. Tough when dried.

Similar Species Outer surface of the Blue Staining Cup (*Caloscypha fulgens*) has blue-green tint.

Habitat In lawns and gardens or other locales where the soil has been disturbed, such as along logging roads and heavily trafficked roadsides. Occurs singly, in close groups, or in clusters.

Range Widely distributed throughout North America.

Scarlet Cup *Sarcoscypha coccinea*

Mushroom enthusiasts living in eastern North America recognize the Scarlet Cup as a harbinger of spring. In the Far West, however, it is a winter mushroom. Despite its bright red color, it is not always readily visible as it may be covered by leaves or partially melted snow. In such instances, its accidental discovery can be a startling experience. The mature mushroom may last for several weeks.

Identification	Saucer shaped to cup shaped, from 1–2½″ wide, with incurved rim. Inner surface of cup scarlet; outer surface whitish with tiny hairs. Stalk usually absent, but when present it is short and white.
Edibility	Unknown; reportedly tastes pleasant.
Similar Species	The Stalked Scarlet Cup (*S. occidentalis*) is smaller and has a prominent white stalk. The Hairy Rubber Cup (*Galiella rufa*) is reddish brown.
Habitat	On twigs and branches lying on the ground, usually partially covered. Occurs singly or in small groups.
Range	Eastern North America; West Coast.

Apricot Jelly *Phlogiotis helvelloides*

This pretty little species is a member of an unusual aggregation known as the jelly fungi. Its texture falls somewhere between rubber and gelatin, and it is one of the few mushrooms that may be eaten raw, usually in salads. It is also used to make a kind of mushroom candy.

Identification	Irregularly funnel shaped or folded. Up to 3″ tall, pinkish-red or salmon color, fading with age. Surface smooth. Stalk short and eccentric.
Edibility	Edible.
Similar Species	The Cinnabar-red Chanterelle (*Cantharellus cinnabarinus*) is not smooth; has gill-like ridges with blunt edges.
Habitat	On rotting wood or on the ground; usually under conifers, but also found in mixed woods.
Range	Widely distributed; most common in the Pacific Northwest and in the Rocky Mountains.

Chanterelle *Cantharellus cibarius*

Much sought after for its delicious, peppery taste, the Chanterelle is equally popular in Europe and America. Its flavor varies somewhat in different areas, but all forms are edible. The family to which the Chanterelle belongs includes many other popular edible species. All have spores on a ridged, wrinkled, or smooth underside.

Identification Cap from 1–6″ wide, sometimes much wider; varies with locale. Pale yellow to orange. Edge of cap inrolled at first, becoming uplifted with age; cap flat to convex, becoming depressed at center. Blunt, forked gill-like ridges, usually paler than cap, run down the stalk in a loose network. Stalk stout to 3″ tall, yellow to whitish, usually smooth. May smell like apricot.

Edibility Choice.

Similar Species Toxic *Omphalotus illudens* has well-formed gills; may have unpleasant odor; on stumps or buried wood. White Chanterelle (*C. subalbidus*) bruises orange.

Habitat On the ground under oaks or conifers; in mixed woods.

Range Throughout North America.

Trooping Cordyceps *Cordyceps militaris*

It is not only humans who are victims of mushroom poisoning. Certain fungi are known to infect grains, other plant forms, and even insects. The Trooping Cordyceps is one such; it grows from the underground larvae and pupae of certain butterflies and moths.

Identification Mushroom to 2″ tall, ¼″ thick. Club-shaped stalk, with wider (upper) portion minutely roughened and bright orange or reddish orange; lower portion smooth and yellow-orange or paler.

Edibility Unknown.

Similar Species Yellow Beetle Cordyceps *(C. melolonthae)* grows from underground large beetle grub. Orange Earth Tongue *(Microglossum rufum)* is smooth above, roughened below. *Mitrula elegans* smooth above; grows in wet soil or water. Irregular Earth Tongue *(Neolecta irregularis)* is smooth and contorted above. Small, smooth *Clavariadelphus* species may be wrinkled.

Habitat Attached to insects buried in the ground or in wood.

Range Widely distributed across North America.

Carmine Coral *Ramaria araiospora*

This beautiful species is one of a large group of mushrooms that resemble a familiar type of undersea coral—hence the genus name, *Ramaria*, meaning "branched." With few exceptions it is easy to recognize a member of the genus in the field. Coral fungi occur in many colors and are widely collected for food. A few, however, are bitter, and some are toxic.

Identification	Coral-red and much-branched; to 5½″ tall, and 3½″ wide. Red overall at first, with branch tips turning yellow. Stalk, when present, is yellow or white.
Edibility	Edible but not recommended; closely resembles at least 1 toxic species.
Similar Species	Toxic Yellow-tipped Coral (*R. formosa*) is pinkish orange to salmon with yellow tips; bruises brown on handling. *R. subbotrytis* bright coral-pink at first, fading to yellowish.
Habitat	On the ground in fall under western hemlock.
Range	From the Pacific Northwest and adjacent Canada to California. Also reported in New York.

Dog Stinkhorn *Mutinus caninus*

The Dog Stinkhorn belongs to a large assemblage of fungi that are related but exhibit amazing structural diversity. A few, like this species, have a columnar form; stinkhorns generally have an odor ranging from merely unpleasant to downright repugnant. They are clearly inedible at maturity, although it is said that young specimens have been consumed.

Identification   Mushroom to 4½″ tall, ⅝″ thick; narrowly cylindrical, spongy, tapering abruptly to a blunt, usually perforated, tip; red, pinkish red, or orange-red, fading to almost white. Tapered area covered by a malodorous, olive-brown slime. Stalk base enveloped by a white sheath.

Edibility   Inedible.

Similar Species   Larger Elegant Stinkhorn *(M. elegans)* is more smoothly tapered, usually more rosy-hued.

Habitat   Leafy debris; near rotting hardwoods; mulch; wood chips.

Range   East of the Rocky Mountains; abundant southward.

Purple Club Coral *Clavaria purpurea*

Coral mushrooms were once all assigned to the genus *Ramaria*; to accommodate structural differences and for other technical reasons, some have recently been classed in the genus *Clavaria*. Although the Latin name means "club-shaped," many members of the genus resemble long, narrow cylinders with pointed or rounded tips.

Identification Clusters of purple cylinders, each to 5″ tall and ¼″ wide; with pointed or blunt tips and white hairs at the base. Color fades to dull brown or paler with age.

Edibility Edible but uninteresting.

Similar Species *C. aurantio-cinnabarina* is deep orange-red. *C. amethystinoides* is paler and sparsely branched. Violet-branched Coral (*C. amethystina*) and *C. zollingeri* are branched; the branches less crowded and more brittle in the latter.

Habitat Under conifers, mostly spruce and fir; in mixed woods.

Range Mountainous areas of western North America.

Hen of the Woods *Grifola frondosa*

The common name of this species is very apt: the fungus frequently resembles a brooding hen, with the overlapping mushroom caps suggesting the bird's plumage. This prime edible sometimes is found growing at the same site year after year.

Identification   Many closely overlapping caps, each ¾–2¾" wide, stemming from a large, branching stalk. Caps spoon-shaped and gray-brown above, white beneath; upper surface smooth to finely hairy, undersurface has fine pores. Stalk white, lateral. Does not stain black with handling.

Edibility   Choice.

Similar Species   *G. umbellata* has centrally stalked caps. The Black-staining Polypore (*Meripilus giganteous*) bruises black on handling.

Habitat   Usually in the soil surrounding the base of oak trees, but near other trees as well, including conifers. Also occasionally on stumps.

Range   Canada to Louisiana; also Midwest and Northwest.

Morel *Morchella esculenta*

This species—also called the Yellow Morel, Blonde Morel, Honeycomb Morel, and Sponge Mushroom—is among the most sought-after of all mushrooms. The related Black Morels *(Morchella angusticeps, M. elata, M. conica)* are also considered choice edibles, but they may cause gastrointestinal problems, especially if consumed with alcohol.

Identification
About 5″ tall, occasionally larger or smaller; cap elliptic, 3″ long, yellow-brown, ridged and pitted, and hollow, fused at its base to the whitish, hollow stalk.

Edibility
Considered choice; some individuals cannot tolerate it.

Similar Species
False Morels (species of *Gyromitra*), some of which are dangerously toxic, have lobed, brain-shaped, chambered caps.

Habitat
Old apple orchards, burned-over areas, hardwood forests, and near dead elms; some types of morels found under conifers.

Range
Widely distributed throughout North America.

Rosy Larch Bolete *Fuscoboletinus ochraceoroseus*

Although bolete-type mushrooms are normally easily distinguished by the undersurface of the cap, this beautiful species seems to occupy an intermediate position, somewhere between boletes (mushrooms with pores) and agarics (mushrooms with gills). Nonetheless, it is unquestionably a bolete, sharing many of their characteristics.

Identification	Cap from 3–10″ broad; dry, hairy to scaly; rose-red to pinkish, usually with yellow edge. Lower surface yellow to brown with long pores and gill-like ridges. Cap edge may have cobwebby veil fragments. Stalk, to 2″, short, stout; yellow and netted above; may have red base.
Edibility	Regarded by some as edible, but usually bitter.
Similar Species	Related *F. paluster* smaller; deeper red; and found in boggy areas. Painted Suillus (*Suillus pictus*), under white pine, and Western Painted Suillus (*S. lakei*), under western conifers, differ in color of spores.
Habitat	Under or near western larch.
Range	Washington, Oregon, Idaho, and adjacent Canada.

Black Helvella *Helvella lacunosa*

Some mushrooms never seem to assume the same form twice, and the Black Helvella is one of them. It is an extremely variable fungus, occurring in a wide range of both shapes and colors. In northern California, when conditions allow, this species may grow to twice its normal size. Although reported to be edible, this species should be avoided; it has many poisonous look-alikes and may itself develop toxic properties with age.

Identification | Cap typically 1–2″ wide; ranges from irregularly saddle-shaped to a series of folds, lumps, or contortions. Dark gray, varying lighter or darker. Stalk pale gray, becoming darker with age, and deeply incised with elongated pits, giving it a fluted appearance.

Edibility | Edible but not recommended.

Similar Species | Resembles other variable fungi, including some toxic ones (notably *Gyromitra* spp.).

Habitat | In grassy areas and in coniferous woods.

Range | Across North America; most abundant on the Pacific Coast.

164

Oyster Mushroom *Pleurotus ostreatus*

The Oyster Mushroom is a very popular edible with a common name that is widely misunderstood. It refers not to the mushroom's taste or texture but to the shape of its cap. Since this species normally fruits abundantly and could supply many meals, hunters and hikers frequently learn to recognize it in case they are lost in the woods. Oyster Mushrooms must be checked closely for beetle infestation.

Identification Caps 2–12″ wide; white or grayish brown, usually in overlapping tiers, with a pleasant odor. Closely set to widely separated gills are white and run down the stalk. Stalk ¼–⅜″; stout, white, and lateral; not always present.

Edibility Choice.

Similar Species Angel's Wings *(Pleurcybella porrigens)* smaller, thin, odorless; found on conifers. *Lentinus* species have sawtooth gill edges.

Habitat On hardwoods, decaying or living. Rarely on pine.

Range Widely distributed in North America.

Honey Mushroom *Armillaria mellea*

Honey Mushrooms comprise a complex of species, varying in size and color but usually edible in all forms. Unfortunately, there are several toxic look-alikes, including one deadly species. Honey Mushrooms grow equally well on living or decaying wood.

Identification Cap 1–4 ½″ wide; yellow, honey-yellow, red-brown, or brown with erect hairs or small scales at center; dry to sticky. Well-spaced gills, white or yellow and stained rusty, extend partway down stalk. Stalk to 6″; whitish, stained with white ring.

Edibility Choice, with caution; some people cannot tolerate it.

Similar Species *A. (Armillariella) tabescens* and the toxic *Omphalotus illudens* lack a ring. Toxic *Gymnopilus spectabilis* and suspect *G. sapineus* are orange-yellow with a yellow ring. Deadly *Galerina autumnalis* smaller; has smooth cap; ring fades. Poisonous *Naematoloma fasciculare* smaller; cap smooth; gills greenish yellow when young.

Habitat Clustered at bases of trees, near stumps, and on soil.

Range Widely distributed across North America.

Mock Oyster *Phyllotopsis nidulans*

There seems little likelihood that the Mock Oyster would be mistaken for the Oyster Mushroom even though they have similar appearances. Among other differences, the Mock Oyster has a very unpleasant odor, whereas that of *Pleurotus ostreatus* is very agreeable and fruity.

Identification Cap 1–3″ wide; orange or yellow-orange, fading paler, and covered with white hairs. Gills closely set (occasionally further apart), and orange or paler. Usually stalkless and in overlapping clusters with caps laterally attached. Odor unpleasant.

Edibility Inedible. The taste is very disagreeable.

Similar Species The Late Fall Oyster *(Panellus serotinus)* has orange gills but is darker and lacks white hairs on the cap surface.

Habitat On logs, stumps, and decaying trees, both coniferous and deciduous.

Range Widely distributed in North America, but more common in the East.

Chicken Mushroom *Laetiporus sulphureus*

Also known as the Sulfur Shelf, this distinctive-looking and brightly colored species has long been prized as an excellent edible. Nevertheless, some people report that it causes stomach upsets as well as other unpleasant experiences. Allergies may be involved here, but some experts suggest that certain of the substrates (the trees upon which the mushrooms grow) may be responsible.

Identification	Large, fan-shaped caps, to 12″ wide, fused at base. Upper surface usually bright orange and wrinkled; color fades to yellow or white. Lower surface yellow with many pores. Stalkless.
Edibility	Choice; may cause sickness in some individuals.
Similar Species	No other mushroom combines orange color, fan-shaped caps, and stalkless growth habit.
Habitat	On a wide variety of deciduous and coniferous trees, both decaying and living. Also on stumps, and on the ground above buried roots.
Range	Widely distributed throughout North America.

Velvet Foot *Flammulina velutipes*

The Velvet Foot is also commonly called the Winter Mushroom, since it can be found fruiting during winter months when thaws permit. It is frequently found in clusters on standing, decaying elms, struck down by the catastrophic Dutch elm disease. A cultivated form of this species is sold commercially as Enotake.

Identification · Cap 1–2½"; reddish yellow, reddish brown, or yellow, with a long, low, central bump; smooth and somewhat sticky. Gills, cream-colored to yellowish, are closely set or wider apart. Stalk to 3" long; narrow; yellowish above, with short, blackish-brown hairs below.

Edibility · Edible.

Similar Species · Toxic Deadly Galerina *(Galerina autumnalis)* has a ring and rusty-brown gills at maturity. The Little Gym *(Gymnopilus penetrans)* has a fragile white tissue over young gills but a ringless stalk.

Habitat · Clustered on logs and on upright, decaying hardwoods, especially American elm.

Range · Widely distributed across North America.

Northern Tooth *Climacodon septentrionale*

This large and rather attractive fungus grows neatly from the side of a standing tree. Its presence there, however, is a sure indication of serious trouble; for this mushroom rots the heartwood of the host tree, doing great damage. The Northern Tooth is frequently found in stands of sugar maple in the Northeast. It is also called *Steccherinum septentrionale*.

Identification
A symmetrical cluster of closely set, shelving caps forming an ellipse up to 12″ tall or taller, with smallest caps at the top and bottom. Each cap buff above, aging to yellow-brown. The lower surface white and covered with crowded, spinelike growths. Dried sections smell like ham.

Edibility
Inedible.

Similar Species
Some polypores similar but have pores, not spines, beneath.

Habitat
Living hardwoods; often on sugar maple.

Range
Northern and eastern U.S. and lower Canada.

Guide to Families

Mushroom genera are said to belong to larger groups, known as families, based on a range of visible or microscopic similarities. The general characteristics of mushroom families are outlined below.

Sarcoscyphaceae

This family and the next three make up the cup fungi. Some Sarcoscyphaceae are like the Scarlet Cup, some are stalked, and shapes vary. Spores are colorless.

Pyronemataceae

Fungi in this family have certain chemical affinities and other technical similarities. Most are stalkless with colorless spores, like the Orange Peel.

Morchellaceae

The most familiar mushrooms in this group, the morels, have shapes suggestive of stalked sponges. The spores are colorless, pale yellow, or pale brown.

Helvellaceae

In general, this group contains the false morels *(Gyromitra)* and similar mushrooms with lobed, folded, or wrinkled caps. Spores are colorless.

Tremellaceae

Many of the jelly mushrooms are gelatinous, but some are tough; they may be lumpy, branched, or funnel-shaped like the Apricot Jelly. Spores are colorless, white, yellow, yellow-orange, or yellow-brown.

Polyporaceae	Most polypores are stalkless and woody, with a pored underside; the vast family includes the edible Chicken Mushroom and Hen of the Woods. Spores are colorless, white, brown, yellow-pink, salmon, or pale blue.
Hydnaceae	Typical tooth fungi have conical spines on the cap underside. A few are edible—not the Northern Tooth or Scaly Hydnum. Spores are white or brown.
Clavariaceae	Many members of this family resemble corals; two are featured here, the Carmine and Purple Club corals. Spores are white, cream, yellow, orange, or brown.
Cantharellaceae	Not all family members measure up to the choice Chanterelle. They bear spores on a ridged, wrinkled, veined, or smooth underside. Spores are white, buff, pink, yellow-pink, yellow, or yellow-orange.
Boletaceae	The soft-textured boletes have centrally stalked caps with a detachable pored layer beneath. Spores are typically olive-brown but may be yellow, pinkish, yellow-brown, reddish brown, or black.
Paxillaceae	These fungi are betwixt and between: they have gills, and so are allied to gilled mushrooms; but the gill layer

detaches, like the pores of the boletes. Spores are brownish, yellowish, olive-green, creamy, or white.

Russulaceae
This family includes two large genera of gilled fungi. *Russula* are extremely fragile and crumble easily. The firmer *Lactarii* release a fluid when if cut. Many of these mushrooms are brightly colored. Spores are white, cream, or yellow.

Hygrophoraceae
Members of this family have bright caps and waxy, sharp-edged gills. The spores in deposit are white.

Amanitaceae
Consisting of the genera *Amanita* and the rarer *Limacella*, this family contains many deadly species. All *Amanita* are initially enclosed in an ovoid sheath of tissue. Spores are white or creamy.

Lepiotaceae
Members of this family have free gills, a ring on the stalk, and scaly caps; some are choice, others toxic. Spores are white, creamy, or greenish.

Tricholomataceae
This vast family includes many widely disparate genera, grouped here on technical grounds. Species in the family may be generally said to meet the following criteria: the spores are white, pale violet, buff, creamy,

ocher, or pinkish; the gills are attached to the stalk and not waxy; the texture is not brittle; and the mushroom does not release fluid when cut.

Volvariaceae — Members of this family have free gills that turn pink with age and pinkish spores. Many grow on wood.

Entolomataceae — These fungi have attached gills that may descend the stalk part way, and pinkish spores. Many are toxic.

Gomphidiaceae — In these mushrooms, the gills are usually well spaced and run down the stalk. Almost all have sticky caps, and the spores are dark gray to black.

Strophariaceae — The genera *Stropharia*, *Psilocybe*, and *Naematoloma* are included here; some experts also include *Pholiota*. Family members have attached gills and purple-brown, yellow-brown, gray-brown, or brown spores.

Cortinariaceae — This large family includes mushrooms with attached gills and rusty, orange-brown, or gray-brown spores. *Cortinarius* species have a webby veil when young.

Coprinaceae — Mushrooms are assigned to this family on technical evidence. In general, family members have attached gills and dark purple-brown or black spores.

Agaricaceae	Typical agarics have a ring on the stalk; the free gills are white, pinkish, or pale gray, turning dark as the chocolate- or purple-brown spores age.
Phallaceae	The aptly named stinkhorns are spongy, columnar fungi with a slimy, deep green tip and a saclike structure at the base. Spores are colorless.
Lycoperdaceae	The puffballs are spherical to pear-shaped, and relatively thin-skinned; they can be as small as a marble or as big as a basketball. Spores are brown.
Geastraceae	The earthstars have an outer skin that folds back at maturity, leaving a star-shaped structure. The brown or purple-brown spores are borne in an inner sphere.
Calostomataceae	Most of the stalked puffballs have a tough or scaly stalks, but the Stalked Puffball-in-Aspic is atypical. Spores are brown to blackish-brown.
Nidulariaceae	The delightful bird's-nest fungi look like diminutive nests to any imaginative person. The white, brown, gray, or blackish spores are enclosed in the "eggs."
Cordycipitaceae	These strangely fascinating fungi grow parasitically upon insects, spiders, or even other mushrooms. The spores are colorless or white.

Spore-bearing Areas of Mushrooms

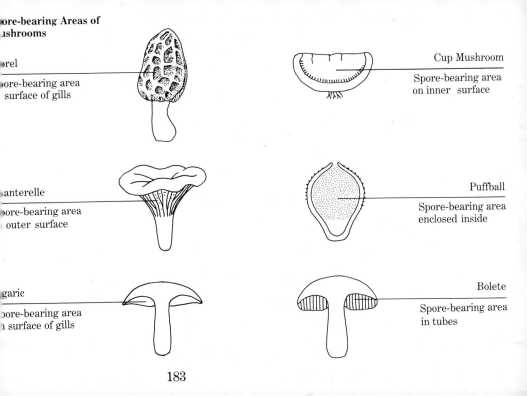

Morel

Spore-bearing area
surface of gills

Chanterelle

Spore-bearing area
outer surface

Agaric

Spore-bearing area
on surface of gills

Cup Mushroom

Spore-bearing area
on inner surface

Puffball

Spore-bearing area
enclosed inside

Bolete

Spore-bearing area
in tubes

Spore Colors in Deposit

How to Make a Spore Print

A spore print is essential for accurate identification of many mushrooms. To make one, cut off the mushroom's stalk close to the base. Place the cap, with the gills or pores facing down, on a piece of white paper. If you are in the field, enclose the cap and paper in wax paper and place them on the bottom of a basket. At home, cover them with a glass. Sometimes the spores fall more readily if you place a drop of water on the cap before you cover it. Some mushrooms produce spore prints in a few hours; others take much longer, sometimes overnight.

White to Cream or Yellow Print

Almond-scented Russula: *pale orange-yellow*
Apricot Jelly: *white*
Aspic Puffball: *pale yellow*
Bleeding Mycena: *white*
Bluing Bolete: *yellow*
Blusher: *white*
Carmine Coral: *yellowish*
Chanterelle: *pale buff to yellowish*
Chicken Mushroom: *white*
Clustered Collybia: *white*
Delicious Lactarius: *buff*
Destroying Angel: *white*
Fairy Ring Mushroom: *white to buff*
Fly Agaric: *white*
Golden Trumpets: *buff*
Gray Amanitopsis: *white*
Hen of the Woods: *white*
Honey Mushroom: *white*
Indigo Lactarius: *yellow to orange-yellow*
Jack O'Lantern: *pale cream*
Larch Waxy Cap: *white*
Man on Horseback: *white*
Northern Tooth: *white*
Orange Mycena: *white*

Oyster Mushroom: *white to lilac-gray*
Parasol: *white*
Parrot Mushroom: *white*
Poison Powderpuff: *white to cream*
Purple Club Coral: *white*
Purple-gilled Laccaria: *white to pale violet*
Rooting Collybia: *white*
Saffron Parasol: *white*
Sandy Laccaria: *white*
Sickener: *white to yellowish white*
Tawny Milkcap: *white*
Trooping Cordyceps: *white*
Velvet Foot: *white*
Yellow Waxy Cap: *white*
White Matsutake: *white*

ink to Salmon, Brownish Pink, r Reddish Print

borted Entoloma: *salmon-pink*
nise Clitocybe: *pinkish cream*
lewit: *pinkish buff*
eer Mushroom: *salmon to brownish pink*
ock Oyster: *pinkish*
ed Chanterelle: *pink to pinkish cream*
ellow Unicorn: *salmon-pink*

urple-brown to Purple-black, moky Black, or Black Print

aymaker's Mushroom: *dark purple-brown*
iberty Cap: *purple-brown*
lica Cap: *brownish black to black*
ld Man of the Woods: *black*
igskin Poison Puffball: *deep brown to violet-black*
osy Gomphidius: *black*
haggy Mane: *black*
ulfur Tuft: *purple-brown*
he Prince: *chocolate-brown to purple-brown*
ine-cap Stropharia: *deep purple-brown*

Ochre- to Rust- or Chocolate-Brown Print

Bracelet Cortinarius: *rust-brown*
Clustered Psathyrella: *dark brown to purple-brown*
Collared Earthstar: *brown*
Deadly Galerina: *rust-brown*
Gilled Bolete: *yellow-brown to olive-green*
Lilac Inocybe: *brown*
Meadow Mushroom: *dark chocolate-brown*
Orange-capped Bolete: *yellow-brown*
Poison Paxillus: *clay-brown to yellowish brown*
Poison Pie: *brown to pale rust-brown*
Questionable Stropharia: *dark purple-brown*
Rosy Larch Bolete: *reddish brown*
Scaly Hydnum: *dull brown*
Scaly Pholiota: *brown*
Violet Cortinarius: *rust-brown*

Gray-green to Olive Print

Frost's Bolete: *olive-brown*
Gem-studded Puffball: *olive-brown*
King Bolete: *olive-brown*

Hyaline (transparent)

Black Helvella
Dog Stinkhorn: *Spores embedded in olive-brown slime.*
Morel
Orange Peel
Scarlet Cup

Glossary

Agaric
A mushroom bearing gills on the undersurface of its cap.

Bolete
A fleshy mushroom bearing a tubelike layer on the undersurface of its cap and belonging to the family Boletaceae.

Bruising
Changing color when handled.

Cap
The top or head of a mushroom.

Cluster
A group of mushrooms rising together from the same spot, typically touching and often attached at the base.

Coniferous
Cone-bearing.

Cup
The saclike tissue at the stalk base of some amanitas, left by the universal veil after it has ruptured.

Depressed
Sunken.

Egg
The immature button stage of amanitas and stinkhorns; one of the spore sacs in a bird's-nest fungus.

Fairy ring
An arc or circle of gilled mushrooms or puffballs.

Free
Not attached to the stalk; used in reference to gills.

Fruiting body
The reproductive portion of a fungus; typically appearing above ground.

Fungus
An organism, traditionally included in the plant kingdom, that lacks chlorophyll and possesses spores.

Gill
One of the radial, bladelike plates that bear spores, located on the undersurface of the cap of many mushrooms.

tex
lear, milky, or colored liquid
t exudes from cut surfaces,
ecially in the genus *Lactarius*.

rgin
e edge of a mushroom cap.

shroom
e fruiting body of a fungus.

cology
e scientific study of fungi.

corrhiza
ymbiotic association between a
gus and the root ends of a
vering plant (plural,
corrhizae).

rasitic
ing in or on another animal or
nt and deriving food from it.

re
e mouth or opening of the tube
ere spores are produced in
etes.

Psychotropic
Having an altering effect on the
mind.

Ring
The remnants left by a partial veil
after it has ruptured; located on
the stalk.

Scale
A piece of the cap of stalk surface
that looks like a shingle.

Spore
The reproductive unit in a fungus.

Spore print
The pattern made by the spores as
they are discharged from gills or
tubes.

Stalk
The portion of a mushroom that
supports the cap and elevates it
sufficiently for adequate spore
dispersal.

Toadstool
Popular term for a poisonous
mushroom.

Veil
A tissue that covers and protects
the immature stage of some gilled
mushrooms and boletes; called a
universal veil when it encloses the
entire immature mushroom and a
partial veil when it covers only the
gills or tubes.

Index

189

The Audubon Society

The NATIONAL AUDUBON SOCIETY, incorporated in 1905, is one of the oldest and largest conservation organizations in the world. Named after American wildlife artist and naturalist, John James Audubon, the Society has nearly 600,000 members in 500 chapters, nine regional and five state offices, as well as a government affairs center in Washington, D.C. Its headquarters are in New York City.

The Society works on behalf of our natural heritage through scientific research, environmental education, and conservation action. It maintains a network of almost 90 wildlife sanctuaries nationwide, conducts ecology camps for adults, and youth programs for schoolchildren. The Society publishes the leading conservation and nature magazine, *Audubon;* an ornithological journal, *American Birds;* and World of Audubon Television Specials, newsletters, video cassettes and interactive discs, and other educational materials.

For further information regarding membership in the Society, write to the NATIONAL AUDUBON SOCIETY, 950 Third Avenue, New York, N.Y. 10022.